ENERGY SCIENCE, ENGINEERING AND TECHNOLOGY

ADVANCED BATTERY PERFORMANCE

SELECT ANALYSES

ENERGY SCIENCE, ENGINEERING AND TECHNOLOGY

Additional books in this series can be found on Nova's website
under the Series tab.

Additional E-books in this series can be found on Nova's website
under the E-books tab.

ENERGY SCIENCE, ENGINEERING AND TECHNOLOGY

ADVANCED BATTERY PERFORMANCE

SELECT ANALYSES

CHRISTOPHER J. MANWELL
EDITOR

Nova Science Publishers, Inc.
New York

NOTICE TO THE READER

The Publisher has taken reasonable care in the preparation of this book, but makes no expressed or implied warranty of any kind and assumes no responsibility for any errors or omissions. No liability is assumed for incidental or consequential damages in connection with or arising out of information contained in this book. The Publisher shall not be liable for any special, consequential, or exemplary damages resulting, in whole or in part, from the readers' use of, or reliance upon, this material. Any parts of this book based on government reports are so indicated and copyright is claimed for those parts to the extent applicable to compilations of such works.

Independent verification should be sought for any data, advice or recommendations contained in this book. In addition, no responsibility is assumed by the publisher for any injury and/or damage to persons or property arising from any methods, products, instructions, ideas or otherwise contained in this publication.

This publication is designed to provide accurate and authoritative information with regard to the subject matter covered herein. It is sold with the clear understanding that the Publisher is not engaged in rendering legal or any other professional services. If legal or any other expert assistance is required, the services of a competent person should be sought. FROM A DECLARATION OF PARTICIPANTS JOINTLY ADOPTED BY A COMMITTEE OF THE AMERICAN BAR ASSOCIATION AND A COMMITTEE OF PUBLISHERS.

Additional color graphics may be available in the e-book version of this book.

Library of Congress Cataloging-in-Publication Data

Advanced battery performance : select analyses / editor, Christopher J. Manwell.
    p. cm.
Includes bibliographical references and index.
ISBN 978-1-61470-456-0 (softcover : alk. paper) 1. Electric batteries. I. Manwell, Christopher J.
TK2901.A37 2011
621.31'242--dc23   2011022603

Published by Nova Science Publishers, Inc. † New York

CONTENTS

PREFACE

This book examines the performance parameters of advanced batteries. The Neosonic polymer Li-ion phosphate battery technology has recently entered the battery market. There are a number of conventional Li-ion FePO4 battery manufacturers from the United States, Taiwan and China, but Neosonic has a more unique design with the polymer Li-ion iron-phosphate technology. The most significant limitation of the iron phosphate cathode material is the low energy density. However, as a result of the safety and power performance improvements, the polymer Li-ion iron phosphate battery technology is being evaluated for new market areas including utility energy storage, and other large energy storage applications.

Chapter 1- The performance of the Neosonic polymer Li-ion battery was measured using a number of tests including capacity, capacity as a function of temperature, ohmic resistance, spectral impedance, hybrid pulsed power test, utility partial state of charge (PSOC) pulsed cycle test, and an over-charge/voltage abuse test. The goal of this work was to evaluate the performance of the polymer Li-ion battery technology for utility applications requiring frequent charges and discharges, such as voltage support, frequency regulation, wind farm energy smoothing, and solar photovoltaic energy smoothing. Test results have indicated that the Neosonic polymer Li-ion battery technology can provide power levels up to the $10C_1$ discharge rate with minimal energy loss compared to the 1 h (1C) discharge rate. Two of the three cells used in the utility PSOC pulsed cycle test completed about 12,000 cycles with only a gradual loss in capacity of 10 and 13%. The third cell experienced a 40% loss in capacity at about 11,000 cycles. The DC ohmic resistance and AC spectral impedance measurements also indicate that there were increases in impedance after cycling, especially for the third cell. Cell #3 impedance R_s

increased significantly along with extensive ballooning of the foil pouch. Finally, at a 1C (10 A) charge rate, the over charge/voltage abuse test with cell confinement similar to a multi cell string resulted in the cell venting hot gases at about 45°C 45 minutes into the test. At 104 minutes into the test the cell voltage spiked to the 12 volt limit and continued out to the end of the test at 151 minutes. In summary, the Neosonic cells performed as expected with good cycle-life and safety.

Chapter 2- In this paper the performance of the LiFeBatt Li-ion cell was measured using a number of tests including capacity measurements, capacity as a function of temperature, ohmic resistance, spectral impedance, high power partial state of charge (PSOC) pulsed cycling, pulse power measurements, and an over-charge/voltage abuse test. The goal of this work was to evaluate the performance of the iron phosphate Li-ion battery technology for utility applications requiring frequent charges and discharges, such as voltage support, frequency regulation, and wind farm energy smoothing. Test results have indicated that the LiFeBatt battery technology can function up to a 10C1 discharge rate with minimal energy loss compared to the 1 h discharge rate (1C). The utility PSOC cycle test at up to the 4C1 pulse rate completed 8,394 PSOC pulsed cycles with a gradual loss in capacity of 10 to 15% depending on how the capacity loss is calculated. The majority of the capacity loss occurred during the initial 2,000 cycles, so it is projected that the LiFeBatt should PSOC cycle well beyond 8,394 cycles with less than 20% capacity loss. The DC ohmic resistance and AC spectral impedance measurements also indicate that there were only very small changes after cycling. Finally, at a 1C charge rate, the over charge/voltage abuse test resulted in the cell venting electrolyte at 110 °C after 30 minutes and then open-circuiting at 120 °C with no sparks, fire, or voltage across the cell.

In: Advanced Battery Performance
Editor: Christopher J. Manwell

ISBN: 978-1-61470-456-0
© 2011 Nova Science Publishers, Inc.

Chapter 1

SELECTED TEST RESULTS FROM THE NEOSONIC POLYMER LI-ION BATTERY[*]

Thomas D. Hund and David Ingersoll

ABSTRACT

The performance of the Neosonic polymer Li-ion battery was measured using a number of tests including capacity, capacity as a function of temperature, ohmic resistance, spectral impedance, hybrid pulsed power test, utility partial state of charge (PSOC) pulsed cycle test, and an over-charge/voltage abuse test. The goal of this work was to evaluate the performance of the polymer Li-ion battery technology for utility applications requiring frequent charges and discharges, such as voltage support, frequency regulation, wind farm energy smoothing, and solar photovoltaic energy smoothing. Test results have indicated that the Neosonic polymer Li-ion battery technology can provide power levels up to the 10C1 discharge rate with minimal energy loss compared to the 1 h (1C) discharge rate. Two of the three cells used in the utility PSOC pulsed cycle test completed about 12,000 cycles with only a gradual loss in capacity of 10 and 13%. The third cell experienced a 40% loss in capacity at about 11,000 cycles. The DC ohmic resistance and AC spectral impedance measurements also indicate that there were increases in impedance after cycling, especially for the third cell. Cell #3

---

[*] This is an edited, reformatted and augmented version of a Sandia National Laboratories publication, SAND2010-4862, dated July 2010.

impedance R_s increased significantly along with extensive ballooning of the foil pouch. Finally, at a 1C (10 A) charge rate, the over charge/voltage abuse test with cell confinement similar to a multi cell string resulted in the cell venting hot gases at about 45°C 45 minutes into the test. At 104 minutes into the test the cell voltage spiked to the 12 volt limit and continued out to the end of the test at 151 minutes. In summary, the Neosonic cells performed as expected with good cycle-life and safety.

ACKNOWLEDGMENTS

The author gratefully acknowledges the assistance of David Johnson at Sandia National Labs for providing the over charge/voltage abuse testing measurements and Wes Baca at Sandia National Labs for test setup and configuration. This work is greatly appreciated.

NOMENCLATURE

AC	alternating current
Ah	Amp-hour
C1	battery capacity in Ah at the 1 h rate
DC	direct current
ESR	equivalent series resistance
FePO4	iron phosphate
ΔIchr	change in current on charge
ΔIdch	change in current on discharge
kg	Kilogram
LiFePO4	lithium iron phosphate
OCVchr	open-circuit voltage before charge
OCVdch	open-circuit voltage before discharge
PSOC	partial state of charge
Rchr	charging resistance
Rdch	discharging resistance
SNL	Sandia National Laboratories
SOC	state of charge
Vmin	minimum operational voltage
Vmax	maximum operational voltage
ΔVchr	change in voltage on charge

ΔVdch   change in voltage on discharge
Wh   Watt-hour
W   Watt

1. INTRODUCTION

This work was funded by Neosonic Li-polymer Energy (Zhuhai) Corp. in Zhu Hai City, Guangdong Province, China. The objectives are consistent with the DOE energy program goals in the following areas:

- Development and evaluation of integrated electrical energy storage systems;
- Development of batteries, superconducting magnetic electrical energy storage (SMES), flywheels, super capacitors and other advanced energy storage devices;
- Analysis and comparison of technologies and applications; and
- Encouraging program participation by industry, academia, research organizations and regulatory agencies.

The work reported in this paper is part of our effort to characterize the performance parameters of advanced batteries. The Neosonic polymer Li-ion iron phosphate battery technology has recently entered the battery market. There are a number of conventional Li-ion $FePO_4$ battery manufacturers from the United States (*A123 and Valence*), Taiwan (*LiFeBatt*), and China (*AA Portable Power and K2 Energy*), but Neosonic has a more unique design with the polymer Li-ion iron-phosphate technology. The most significant limitation of the iron phosphate cathode material is the lower energy density at approximately 80 Wh/kg vs. 165-180 Wh/kg for $LiCoO_2$ and other oxide cathodes. However, as a result of the safety and power performance improvements, the polymer Li-ion iron phosphate battery technology is being evaluated for new market areas including utility energy storage, and other large energy storage applications. Other advantages of this technology are the thin prismatic form factor, foil pouch packaging, and a gel electrolyte to minimize volatile liquid components while maintaining good performance at low temperatures[1]. In this paper the performance of the Neosonic polymer Li-ion iron phosphate cell was evaluated using high rate capacity tests, low temperature capacity, ohmic resistance, spectral impedance, partial state of charge (PSOC) pulse power cycling, pulse power performance, and an over-charge/voltage abuse test[2].

In Figure 1 is a photograph of one of the prismatic Neosonic polymer Li-ion 10 Ah cell. This cell has dimensions of 101 mm wide by 205 mm long by 8.5 mm thick, and weighs 350 gm. Table 1 summarizes the Neosonic polymer Li-ion cell specifications.

Table 1. Neosonic Li-ion polymer Cell Specifications

Model #KMBNF82100202R Cell	
Operating voltage window	2.10 – 3.65 V
Max voltage	3.65 V
Discharge end voltage (1C rate)	2.10 V
Charge Regulation Voltage	3.65 V
Full Charge Termination (V, I, Time)	3.65 V and 0.15 A or 60 min @ 3.65 V
Maximum Charge Current	30 A
Maximum Pulse Current, I	100 A
Maximum Constant Current, I	50 A
Internal Ohmic resistance, mOhm +25 °C	<3 mohm
Ah Capacity (0.25C and 1C rate)	10,000 mAh,
Energy stored in operating voltage window, Wh or kJ	80 Wh/kg 170 Wh/L
Overall dimensions, mm	101 x 205 x 8.5 mm
Weight, kg	0.350
Operating temperature, °C	0 to 45 Charge -10 to 60 Discharge
Storage temperature, °C	-10 to 45
Cycle life, cycles	

Figure 1. Neosonic Li-ion polymer Li-FePO4 Cell Under Test.

2. TEST PROCEDURES

The cell test procedures used in this effort were initially developed as part of a plan to test batteries for partial state of charge (PSOC) pulsed cycling in utility applications, and the eight characterization tests used are itemized below.

1) **Capacity Test** – Establishes a capacity on each cell.
2) **DC Ohmic Resistance** – Establishes a resistance of the cell.
3) **AC Spectral Impedance** - Establishes the AC impedance of the cell.
4) **Cell Power Density and Specific Energy Density** – Measures the cell power and energy density.
5) **Cell Capacity and Recharge As A Function Of Temperature** – Capacity measurements at the 1C rate were conducted at 35, 22, 0, and -20°C.
6) **Cell Utility PSOC Pulsed Cycle Test** – Measures the ability of the cell to PSOC cycle at high power for utility voltage support, frequency stabilization, and wind farm energy smoothing applications.
7) **Hybrid Pulse Power Test** – Measures the 10 second pulse power performance from 90% to 10% state of charge (SOC).
8) **Over Voltage/Charge Abuse Test** – Measures the effects of an uncontrolled continuous 1C (10 A) charge.

2.1. Capacity Test

The capacity test is used to determine the cell capacity and this test is done prior to testing to establish a baseline as well as being repeated at the end of PSOC cycle testing. This will help identify how the various tests in this plan affect the cell capacity.

The cell shall be tested for its capacity, as follows:

1) Each cell shall be charged at 0.5C (5.0 A) up to Charge Voltage (3.65 V). On reaching the Charge Voltage, the current shall be allowed to taper while maintaining the Charge Voltage until the current tapers to 0.15 A. The ampere-hour input into the cell shall be measured.
2) The cell shall rest at open circuit for 60 minutes.
3) The cell shall be discharged at 0.5C (5.0 A) until the end voltage (2.1 V) is reached. The ampere-hour capacity of the cell shall be measured.

4) The cell shall rest at open circuit for 60 minutes.
5) Steps 1 to 5 shall be repeated 3 times. The third capacity measurement will be the recorded capacity.

2.2. DC Ohmic Resistance

The DC resistance shall be measured with a high current discharge pulse of at least 1C at the 100% SOC point. An oscilloscope will be used to measure the ohmic voltage drop and current ramp using the following steps:

1) Charge cell at 1C rate to the Charge Voltage (3.65 V) and hold at voltage until the current tapers to 0.15 A.
2) Allow the cell to sit open-circuit for 60 min.
3) Measure the dynamic DC ohmic resistance of the cell using an oscilloscope by discharging at 1C for 2 seconds.

2.3. AC Spectral Impedance

AC spectral impedance measurements of the as-received and after PSOC pulsed cycling were also made. Impedance data was collected on the Neosonic cells both before and after cycle life testing, and in one case only after cycling. Prior to making any measurements all cells were brought to 100% SOC, placed into an environmental chamber whose temperature was regulated to 25°C, and allowed to sit at open circuit at least 24 hrs in order to fully equilibrate. All measurements were made using a Solatron SI 1287 potentiostat and model 1255B frequency analyzer using a 4-wire configuration over a frequency range of 1 Mhz to 0.1 mhz. These measurements shall be made in the following manner:

1   Each cell shall be charged at 1C up to the Charge Voltage (3.65 V) until the current tapers to 0.15 A.
2   The cell shall rest at open circuit for at least one day prior to measurement.
3   Each terminal on the cell shall be fitted with two gold plated interconnects (to minimize contact impedance).
4   The cell shall be placed into an environmental chamber regulated at 25°C for at least 12 h prior to initiating measurements.

5   All measurements shall be made with the cell in a controlled temperature maintained at 25°C.
6   The cell shall be connected to the instrumentation in a four-wire configuration.
7   The peak-to-peak AC voltage shall be in a range to allow 1% accuracy of the impedance of the cell being measured, and in this case 3 mV was used.
8   The frequency range shall be large enough to encompass the anticipated network response, and in this case corresponds to a range of 100 kHz to 10^{-4} Hz. At least six different frequencies per decade shall be measured.
9   All measurements shall be made at 0 V vs. the open circuit voltage corresponding to 100% SOC.

2.4. Cell Power Density And Specific Energy

Measure cell capacity as close as possible to the following rates:

1	0.1C1,	10 h
2	0.2C1,	5 h
3	1C,	1 h
4	2C1,	0.5 h
5	4C1,	0.25 h
6	10C1,	0.1 h

The cell capacity shall be measured as follows:

1   Each cell shall be charged at 1C up to Charge Voltage (3.65V). On reaching the Charge Voltage, the current shall be allowed to taper while maintaining the Charge Voltage until the current tapers to 0.15 A. The ampere-hour input into the cell shall be measured.
2   The cell shall rest at open circuit for 30 minutes.
3   The cell shall be discharged at specified rate until the end voltage (2.1 V) is reached. The ampere-hour and Watt-hour capacity of the cell shall be measured.
4   Steps 1 to 4 shall be repeated 3 times.

Using the data above, calculate the power and energy density with respect to volume and weight and display using the Ragone plot.

2.5. Cell Capacity and Recharge as a Function of Temperature

Cell capacity and charge and discharge characteristics at the 1C rate (10 A) were measured at 35, 22, 0, and -20°C. The capacity test procedure was the same as that described in the capacity measurement.

2.6. Utility PSOC Pulsed Cycle Test

The utility PSOC pulsed cycle test is designed to evaluate battery performance under short high power charge and discharge environments. In many utility applications the battery is required to both sink and source power for voltage support, frequency stabilization, and wind farm energy smoothing. In Figure 2 are actual utility data obtained from Charles Koontz of WPS Energy Services, Inc. showing the magnitude and duration of the power pulses required to support a utility application. In general, the pulse durations are minutes in length. The utility PSOC charge and discharge pulses chosen for this test were between 1.5 and 3 minutes in length at discharge rates between 2C1 (20 A) and 4C1 (40 A). The goal of this testing is to evaluate PSOC pulsed cycling, cell stability, efficiency, power performance, thermal management, and charge management strategies.

The cycle profile in this test is illustrated in Figure 3 and consists of the following steps:

1   Charge cell at 1C rate until voltage reaches Charge voltage (3.65 V).
2   Keep voltage at Charge voltage until current tapers to 0.15 A.
3   Rest for 30 min.
4   Discharge at 1C rate to end voltage (2.1 V).
5   Rest for 30 min.
6   Recharge cell as in step 2.
7   Discharge at 1C rate to 50% Ah capacity.
8   Rest for 5 min.
9   Discharge at 2 or 4C1 rate for 3 or 1.5 min. (10% DOD).
10 Rest for 5 min.
11 Charge at 2 or 4C1 rate for 3 or 1.5 min.

12 Rest for 5 min.
13 Repeat steps 9 through 12 for 1,000 cycles.
14 Measure available capacity as specified in steps 3 through 5.
15 Repeat 1,000 cycle profile one to three times per test sequence.
16 Evaluate battery performance and determine if higher or lower power levels are necessary to maintain the PSOC cycle interval of 1,000 cycles.

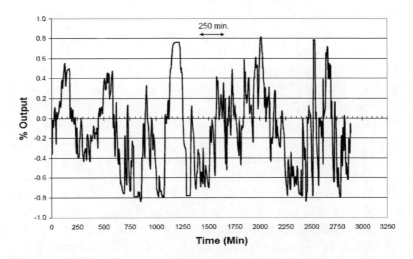

Figure 2. Typical Utility Energy Pulses (Charles Koontz, WPS).

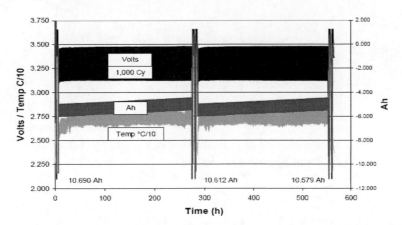

Figure 3. Utility PSOC Pulsed Cycle Test, Cell #1, with 2C1 3 min. (10% DOD) Charge/Discharge Pulses, (Cycle 0 to 2,000).

2.7. Hybrid Pulse Power Test

The Hybrid Pulse Power Test is extracted from the FreedomCAR Battery Test Manual For Power-Assist Hybrid Electric Vehicles. This test procedure uses a 10 second 5C1 discharge pulse and a 3.75C1 charge pulse 40 seconds apart (see Fig. 4). The test sequence is listed below:

1   Measure capacity at the 1C rate.
2   Fully recharge cell.
3   Allow cell to rest open-circuit for 1 h.
4   Discharge cell 10% at the 1C rate,
5   Allow the cell to rest for 1 h rest open-circuit (measure Voc).
6   Discharge cell at the 5C1 rate for 10 seconds (measure end of discharge V).
7   Allow the cell to rest open-circuit for 40 seconds (measure Voc).
8   Charge at the 3.75 C1 rate for 10 seconds (measure end of charge V).
9   Discharge at the 1C rate 10% of the cell capacity.
10 Repeat steps 4 through 8 until battery is at 10% SOC.
11 Record open-circuit voltage after the 1 h rest before the discharge pulse, record voltage at 10 second point in charge and discharge pulse and record open-circuit voltage at end of 40 second rest for each SOC.
12 Calculate discharge resistance using the 1 h open-circuit voltage and charge resistance using the 40 second open-circuit voltage for each SOC.

$$R_{Dch} = \frac{\Delta V_{Dch}}{\Delta I_{Dch}}$$
$$R_{Chr} = \frac{\Delta V_{Chr}}{\Delta I_{Chr}}$$

13 Calculate the Discharge Pulse Power Capability for each SOC using the minimum operational voltage.

$$Watts = V_{min} \bullet (OCV_{Dch} - V_{min}) \div R_{Dch}$$

14 Calculate the Charge Pulse Power Capability for each SOC using the maximum operational voltage.

$$Watts = V_{max} \bullet (V_{max} - OCV_{Chr}) \div R_{Chr}$$

15 Plot the discharge and charge power as a function of % SOC and discharged energy (Wh) at the 1 h rate.

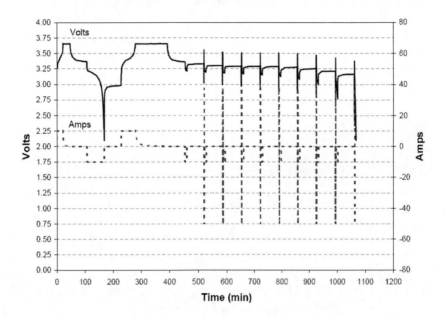

Figure 4. Neosonic Hybrid Pulse Power Test.

Max power was calculated from the recorded data and using the equations shown in steps 13 and 14 above. Using ohms law, current is calculated by dividing the voltage by resistance. The voltage difference between the cell open-circuit and cell max and min voltages divided by resistance is assumed to be a measure of maximum current. In the second half of the two equations, the min and max voltages are multiplied by the calculated current to yield max 10 second power (Watts = V x I). Once the max charge and discharge power is calculated, the power data can be plotted as a function of % SOC and battery energy level in Wh. The power vs. energy plot can then be used to scale an energy storage system for any power and energy requirement at specified pulse durations.

2.8. Over Voltage/Charge Abuse Test

The Over Voltage/Charge Abuse Test was defined by a 1C charge rate (10 A) up to 12 V until the cell failed. The over charge test conducted for this report included mechanical support for the foil pouch cell using a confining top and bottom plate. The confining plate was introduced to more closely simulate a multi-cell battery pack. The charge and data acquisition was terminated when the cell under test lost all voltage and/or exceeded the time limit and/or stabilized at current and voltage.

3. TEST RESULTS

3.1. Initial and Final Capacity

Figure 5 through 7 show the voltage and capacity (Ah) data for a discharge and charge on cell #1, 2, 3 as received and after 11,812, 12,454, and 11,078 PSOC pulse cycles. All initial cell capacities are within the capacity specifications of between 9 and 10 Ah. At end of test cell #1 and 2 capacities are down by 10 and 13% and Cell #3 capacity is down by 40%. Cell #1 and 2 also have a small drop in discharge voltage after pulsed cycling, but cell #3 has a large voltage drop of about 0.20 V on discharge.

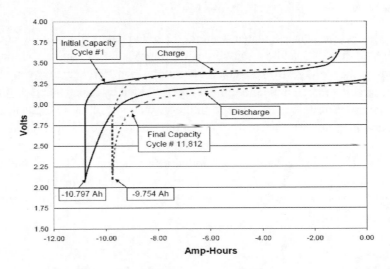

Figure 5. Initial and Final Capacity of Cell #1 (5A Chr/Dch).

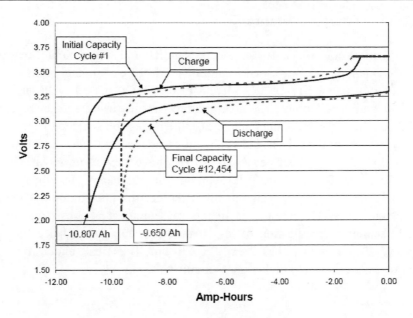

Figure 6. Initial and Final Capacity of Cell #2 (5A Chr/Dch).

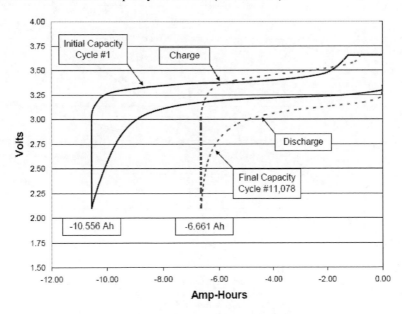

Figure 7. Initial and Final Capacity of Cell #3 (5A Chr/Dch).

3.2. DC Ohmic Resistance

Figure 8 thr ough 10 shows the initial and final ohmic resistance measurement for cell #1, 2, and 3 at 0.0024, 0.0013, and 0.0008 ohms ($\Delta V/\Delta I$) and after cycling at 11,812, 12,454, and 11,078 PSOC pulsed cycles at 0.0068, 0.0040, and 0.0075 ohms. The measurement is made using a linear regression to obtain the slope of the line. In this case the regression is a good fit to data.

The slope of the current and voltage measurement is the average resistance value for a discharge between 0 and -10 A. This battery chemistry has a slope that is somewhat consistent throughout the discharge. In some battery technologies there can be a significant difference from beginning to end of the discharge pulse. The initial data in this measurement is consistent with the manufacturer's specification of less than 0.003 ohms impedance, but after cycling the resistance has clearly gone up. The final ohmic resistance measurement also has a higher open-circuit voltage.

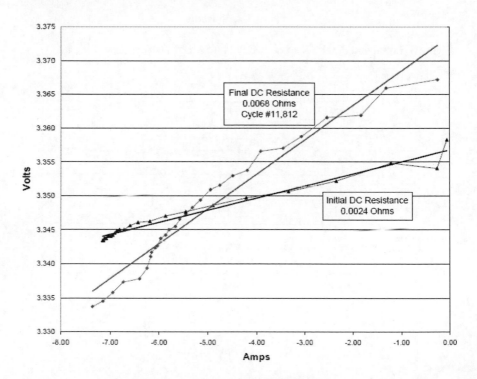

Figure 8. Ohmic Resistance Mcasurement On Cell #1 (100% SOC).

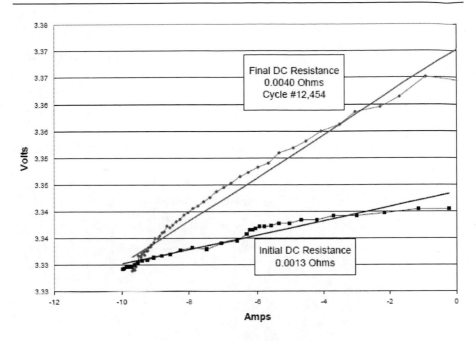

Figure 9. Ohmic Resistance Measurement On Cell #2 (100% SOC).

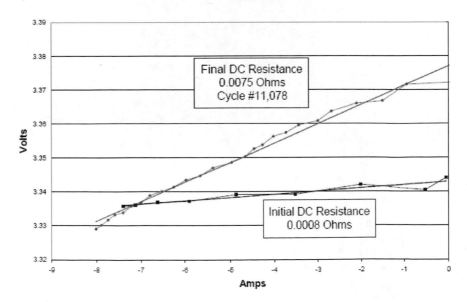

Figure 10. Ohmic Resistance Measurement On Cell #3 (100% SOC).

3.3. AC Spectral Impedance

A Nyquist plot of the impedance data before and after PSOC pulsed cycling is shown in Figure 11 through 14, and as seen, there is a significant increase in impedance after cycling. Shown in Figure 11 are the data collected for the two Neosonic cells for which complete data sets were collected over the entire frequency range, and in Figure 12 are the high frequency regimes of the same data sets. Figure 13 shows the impedance data for all three Neosonic cells after cycling.

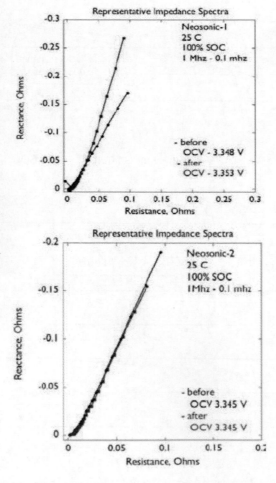

Figure 11. Impedance data of two Neosonic cells before and after cycle-life testing at 25^0C and 100% SOC.

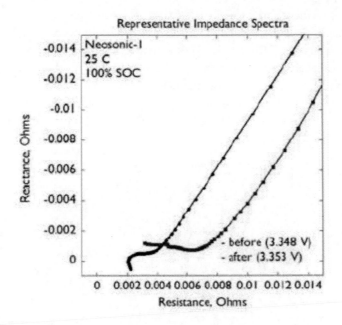

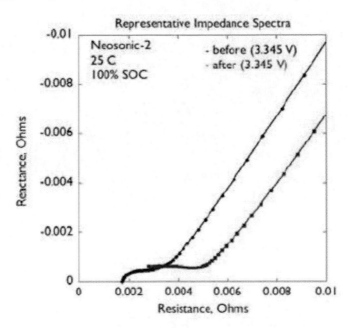

Figure 12. High frequency impedance data of two Neosonic cells before and after cycle-life testing at 25^0C and 100% SOC.

As is evident on casual inspection of the high frequency data, the impedance of the cells is seen to increase after having been cycled. For a more complete analysis of the data the equivalent circuit diagram shown in Figure 14a is a reasonable representation for the cell. However, the lack of sufficient information regarding the anode and cathode and the appearance of the data in which two distinct RC processes are not clearly evident and resolved makes use of the equivalent circuit shown in Figure 14b the preferable alternative.

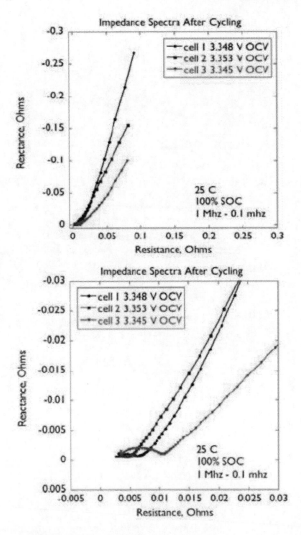

Figure 13. Comparison of impedance data of the three Neosonic cells after cycling. Full frequency range shown on left, and high frequency data shown on right.

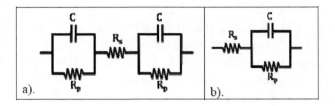

Figure 14. Equivalent circuit for cell (on left) and simplified circuit used for fitting (on right).

Using the equivalent circuit model shown in Figure 14 b and the high frequency impedance data the resistances for the charge transfer (R_p) and electrolyte solution (Rs) resistances for the cells were estimated and are summarized in Table 2. As seen, in the case of cells 1 and 2 for which before and after data sets are available, the behavior is similar. The solution resistance for the cells is seen to increase from 2.2 and 2 mΩ to 4.1 and 4.2 mΩ for cells 1 and 2, respectively. Also as seen, cell 3 exhibits similar after cycling behavior, and has an Rs of 4.2 mΩ.

The charge transfer resistance (R_p) of the cells is also seen to increase. In the case of cell 1 the charge transfer resistance increases from 1.5 to 2.6 mΩ while in the case of cell 2 the resistance goes from 1.6 to 3.6 mΩ. In the case of cell 3 the observed charge transfer resistance is even larger, on the order of 7.5 mΩ.

Table 2. Estimated circuit elements using simplified equivalent circuit

Cell ID	R_s (mΩ)	R_p (mΩ)	$R_s + R_p$ (mΩ)
1			
Before cycling	2.2	1.5	3.7
After cycling	4.1	2.6	6.7
2			
Before cycling	2.0	1.6	3.6
After cycling	4.2	3.6	7.8
3			
After cycling	4.2	7.5	11.7

The instantaneous ohmic drop of the cell corresponds to the electrolyte solution resistance R_s. However, depending on the size of the parallel capacitors (the porous electrodes) and their resultant RC-time constant in combination with the DC-test equipment used for making pulsetype measurements (the switching speeds and sampling rate of the battery cycling equipment), the instantaneous voltage drop for the cell will be estimated as R_s, or some combination of R_s and R_p up to a maximum value of $R_s + R_p$ as shown in Table 2. Regardless, the effective resistance of cells 1 and 2 essentially doubles on cycling, and in the case of cell 3 increases by a factor of approximately three if one assumes that its initial resistance prior to cycling is comparable to cells 1 and 2.

Another interesting aspect of cell 3 behavior was that during the course of 48 hrs over which the impedance measurements were made the cell generated gas, as evidenced by swelling of the cell. The reason for this behavior is not known, but the experimental conditions employed were not outside the performance envelope of the cell, and in fact cells 1 and 2 continued to operate normally after being subjected to the same test protocol with the same equipment.

3.4. Cell Power and Specific Energy Density

In Figure 15 is a plot of the cell #6 specific energy and power performance and in Figure 16 is the energy and power density using the Ragone plot. The results show a steep curve as the power level increased indicating minimal energy losses up to the $10C_1$ (100A) rate.

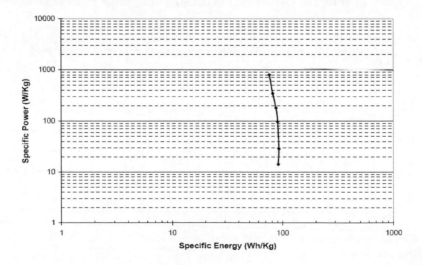

Figure 15. Specific Energy And Power For Cell #6.

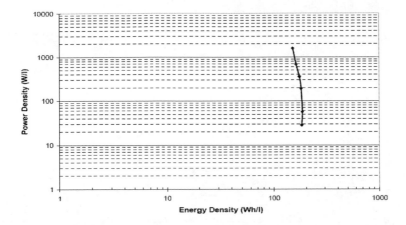

Figure 16. Energy Density And Power Density For Cell #6.

3.5. Cell Capacity and Recharge as a Function of Temperature

In Figure 17 are the 1C capacity measurements at 35, 25, 0, and -20°C. The results show a rapid drop in capacity at -20°C and below, and at -30°C there is no usable capacity. In Figure 18 are the recharge voltage profiles at the 1C rate. At -20°C and below, the recharge is dramatically slowed due to the rapid increase in cell voltage up to the regulation voltage. At these low temperatures the recharge currents are low and require much longer recharge times.

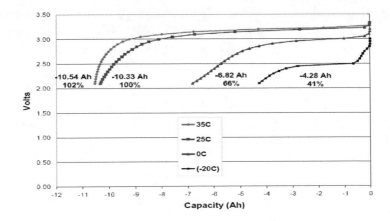

Figure 17. Capacity vs. Temperature at 10 A.

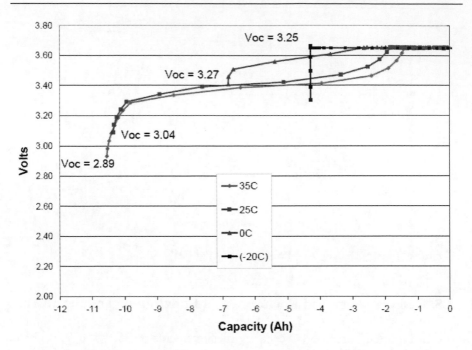

Figure 18. Recharge vs. Temperature at 10 A.

3.6. Hybrid Pulse Power Test

The hybrid pulse power test results on Cell #6 are shown in Figure 19 and 20. As expected, the maximum discharge pulse at a low state of charge is significantly reduced and as the SOC increases so does the discharge pulse power, from 136 W (10% SOC) to 417 W (90% SOC). Usually the charge pulse has a similar decrease in power as the SOC increases, but in this case it remains fairly constant at about 219 W between 30 and 70% SOC. In Figure 20 the available battery energy is plotted on the abscissa (X-axis) providing an operational range with available energy. This can be useful for scaling up to the necessary power and energy levels needed to meet the 10 second pulse power requirements of the desired system.

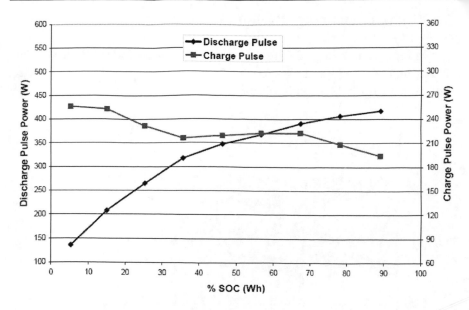

Figure 19. Hybrid Pulse Power Capability As A Function Of %SOC.

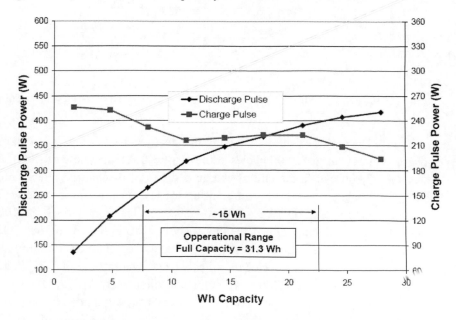

Figure 20. Hybrid Pulse Power Capability As A Function Of Capacity In Wh.

3.7. Utility PSOC Pulse Cycle-Life Test

In Figure 21 through 25 are the utility PSOC pulsed cycle-life test results at $2C_1$ and $4C_1$ (±20 and ±40 A) and the summary capacity plot of all three cells in the Utility PSOC Pulse Cycling Test.

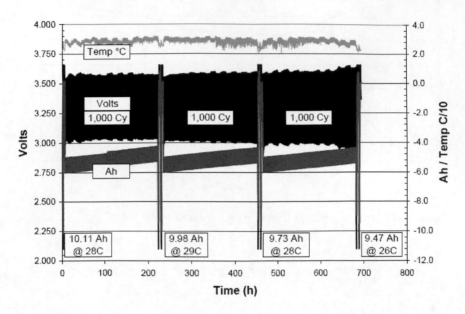

Figure 21. Utility PSOC Pulse Cycle-Life Test Results Cell #2 At ±40A, Cycle 9,000 to 12,000.

Figure 21 is a plot of cell #2 showing the steady end of charge and end of discharge voltage up to 12,000 cycles at the $4C_1$ (±40 A) rate. Between 11,750 cycles and 12,000 cycles the end of charge and end of discharge voltages are beginning to expand for the first time during the test. In this plot the cell end of charge voltage steps up to 3.64 V at 12,000 cycles from 3.53 V at 100 cycles, while the cell temperature remains at about 28°C at the end of the cycle sequence. The end of charge and discharge voltage was relatively stable in all cycle sequences up to 8,694 cycles for all cells. The average end of charge and discharge voltage at cycle 6,500 at 40 amps was about 3.57 V and 3.03 V. In Figure 22 through 25 the last test sequence for cells #1, #2, and #3 are shown. In all cases the end of charge voltage rises to the high voltage limit of 3.65 V after 19, 454, and 66 cycles. As a result of the utility pulsed cycling, the ability of the cell to charge and discharge at the $4C_1$ rate has been significantly degraded starting at about 8,700 for cell #3 and about 12,000 cycles for cells #1 and 2.

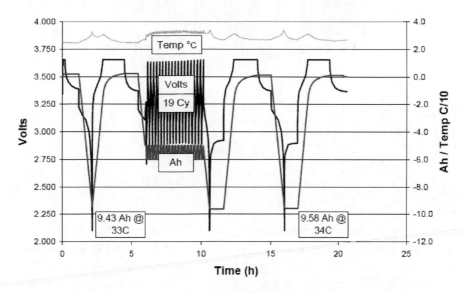

Figure 22. Final Utility PSOC Pulse Cycle-Life Test On Cell #1 At ±40 A, Cycle #11,793 to 11,812.

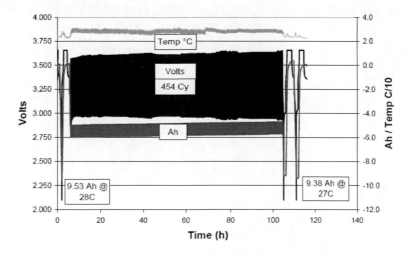

Figure 23. Final Utility PSOC Pulse Cycle-Life Test On Cell #2 At ±40 A, Cycle #12,000 to 12,454.

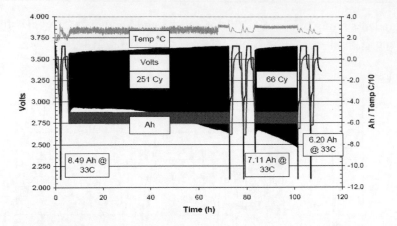

Figure 24. Utility PSOC Pulse Cycle-Life Test On Cell #3 At ±20 A, Cycle #10,761 to 11,078.

Figure 25 is a capacity summary at the 1 hr rate for the three different cells. All cells were cycled at the $2C_1$ rate up to 4,000 cycles, then after that all cells were cycled at the $4C_1$ rate up to 8,761 cycles where only Cell #3 was cycled at the $2C_1$ rate. The results in Figure 25 show that the capacity of cell #3 drops dramatically after about 10,000 cycles, while the other two cell capacities are slowly fading to less than 9.5 Ah out beyond 12,000 cycles. Based on the abrupt capacity loss in cell #3 and the overall trend in capacity, cells #1 and 2 should maintain the slow trend in capacity loss beyond 15,000 cycles.

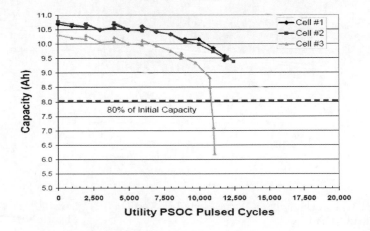

Figure 25. Summary Capacity For Utility PSOC Pulse Cycle Test.

3.8. Over Voltage/Charge Abuse Test

In Figure 26 are the voltage, current, and temperature plots from the over charge/voltage abuse test with cell confinement. Since the cells are sealed in a foil pouch without any mechanical support, the effect of mechanical confinement is to significantly improve the outcome of the overcharge test by minimizing the cell bulging with gas and vented smoke.

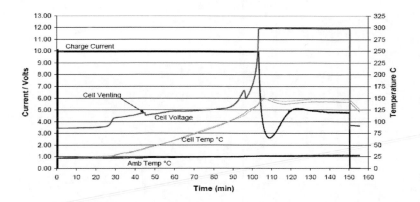

Figure 26. Over Voltage/Charge Abuse Test With Confinement, Cell #5, At 1C (10A) Rate.

In Figure 26 and 27 the cell is confined mechanically with a fiberglass reinforced plate. Figure 26 shows the foil pouch venting at 45 min into the test at a voltage of 4.54 V and at a temperature of 44^0C with no visible smoking during the test. The confined cell voltage and temperature increases slowly until about 100 min. into the test, then the voltage spikes to the 12 V limit, and after about 50 min at 12 V the test is terminated. The max temperature of the confined cell only reaches about 150^0C with no visible smoke, electrolyte, or fire (see Figure 27).

#1 Confinement   #2 Pouch Cell After Test

Figure 27. Photos of Over Voltage/Charge Abuse Test With Confinement Cell #5.

Table 3. Neosonic Summary Test Results

Test	Item	Cell #1	Cell #2	Cell #3	Cell #4	Cell #5	Cell #6
Initial Capacity C/2 (Ah)		10.8	10.8	10.6	10.7	10.7	10.6
Final Capacity Test C/2 (Ah)		9.7	9.7	6.7	NA	NA	NA
Initial DC Ohmic Resistance (Ohms)		0.0024	0.0013	0.0008	NA	NA	NA
Final DC Ohmic Resistance (Ohms)		0.0068	0.004	0.0075	NA	NA	NA
Initial AC Spectral Impedance Rs (Ohms)		0.0022	0.002	NA	NA	NA	NA
Final AC Spectral Impedance Rs (Ohms)		0.0041	0.0042	0.0042	NA	NA	NA
Cell Power and Energy Density at 10C1 Rate	W/l	NA	NA	NA	NA	NA	1,590
	Wh/l						184
	W/Kg						800
	Wh/Kg						92
Cell Capacity As A Function Of Temperature	Ah @ 35°C	NA	NA	NA	NA	NA	10.54
	Ah @ 25°C						10.33
	Ah @ 0°C						6.82
	Ah @ -20°C						4.28

Table 3. (Continued).

Test	Item	Cell #1	Cell #2	Cell #3	Cell #4	Cell #5	Cell #6
Utility Cycle-Life	Test Cycles	11,812	12,454	11,078	NA	NA	NA
	% Cap Loss	10	13	40			
Hybrid Pulse Power Test, Power (W) at 50% SOC	Charge	NA	NA	NA	NA	NA	219
	Discharge						347
Over Voltage/Charge Abuse Test	Vent T °C					44°	
	Max T °C	NA	NA	NA	NA	150°	NA
	Fire Y/N					No	

4. SUMMARY

Overall the Neosonic polymer Li-ion cells cycled well. Capacity degradation was measured on cells #1 through #3 of 10, 13, and 40%, after 11,812, 12,454, and 11,078 cycles as a result of the utility PSOC cycling (see Table 3). Based on the trend of capacity fade for cell #3, cells #1 and #2 may PSOC pulse cycle beyond 15,000 cycles before reaching 80% of their initial capacity. The ohmic resistance measurements and spectral impedance measurements before and after the PSOC pulse cycling have indicated an increase in ohmic resistance and Rs as a result of the cycling. The ohmic value increased from 2.4, 1.3, and 0.8 mohms to 6.8, 4.0, and 7.5 mohms, while the AC spectral impedance Rs value increased from 2.2, and 2.0 to 4.1 and 4.2 mohms. This is a significant increase and indicates degradation in the cell's electrochemical processes as a result of the cycling.

The other three cells were characterized for over voltage/charge, cell pulse power, and capacity vs. temperature, as indicated in Table 3. The cell capacity as a function of temperature at 35, 25, 0, and -20°C show that both discharge capacity and recharge voltage are significantly affected by lower temperatures especially at or below 0°C. From -20°C to 25°C the capacity increased by about 1.3% per °C. The 10 second pulse power capability values measured power levels of 347 W/cell on discharge and 219 W/cell on charge at 50% SOC. Finally, the over charge/voltage abuse test showed that the Neosonic cell can fail safely without fire or damage to other external systems if the cell foil pouch is maintained under compressive pressure in the cell stack. Max cell temperature was only measured at 150°C after gaseous electrolyte venting. These results show that the Neosonic cells as tested meet or exceed the manufacturer's specifications in capacity, internal ohmic resistance, max power, specific energy, and safety.

5. REFERENCES

[1]   Handbook Of Batteries, 3th edition, G. W. Linden, McGraw-Hill Handbooks 1995, ISBN 0-07-135978-8.
[2]   Hund T. D., Ingersoll D., *Selected Test Results from the LiFeBatt Iron Phosphate Li-ion Battery*, Sandia National Laboratories. Report# SAND2008-5583. September 2008. Technical Reports, http://www.sandia.gov/ess/Publications/pubs.html#techreports.

In: Advanced Battery Performance
Editor: Christopher J. Manwell

ISBN: 978-1-61470-456-0
© 2011 Nova Science Publishers, Inc.

Chapter 2

SELECTED TEST RESULTS FROM THE LIFEBATT IRON PHOSPHATE LI-ION BATTERY[*]

Thomas D. Hund and David Ingersoll

ABSTRACT

In this paper the performance of the LiFeBatt Li-ion cell was measured using a number of tests including capacity measurements, capacity as a function of temperature, ohmic resistance, spectral impedance, high power partial state of charge (PSOC) pulsed cycling, pulse power measurements, and an over-charge/voltage abuse test. The goal of this work was to evaluate the performance of the iron phosphate Li-ion battery technology for utility applications requiring frequent charges and discharges, such as voltage support, frequency regulation, and wind farm energy smoothing. Test results have indicated that the LiFeBatt battery technology can function up to a 10C1 discharge rate with minimal energy loss compared to the 1 h discharge rate (1C). The utility PSOC cycle test at up to the 4C1 pulse rate completed 8,394 PSOC pulsed cycles with a gradual loss in capacity of 10 to 15% depending on how the capacity loss is calculated. The majority of the capacity loss occurred during the initial 2,000 cycles, so it is projected that the LiFeBatt should PSOC cycle well beyond 8,394 cycles with less than 20% capacity loss.

---

[*] This is an edited, reformatted and augmented version of a Sandia National Laboratories publication, dated September 2008.

The DC ohmic resistance and AC spectral impedance measurements also indicate that there were only very small changes after cycling. Finally, at a 1C charge rate, the over charge/voltage abuse test resulted in the cell venting electrolyte at 110 °C after 30 minutes and then open-circuiting at 120 °C with no sparks, fire, or voltage across the cell.

ACKNOWLEDGMENTS

The author gratefully acknowledges the assistance of Michelle Robinson and Don Harmon of LiFeBatt, Inc. USA, and in addition Alan Hsu and Dr. Jerry Yao (Dr.Yao owns the license from ITRI Taiwan) of LiFeBatt Taiwan for providing the Li-ion Iron phosphate cells. Also, the assistance of David Johnson at Sandia National Labs for providing the over charge/voltage abuse testing measurements and Wes Baca at Sandia National labs for test setup and configuration is greatly appreciated.

NOMENCLATURE

AC	alternating current
Ah	Amp-hour
C1	battery capacity in Ah at the 1 h rate
DC	direct current
ESR	equivalent series resistance
FePO4	iron phosphate
ΔIchr	change in current on charge
ΔIdch	change in current on discharge
kg	Kilogram
LiFePO4	lithium iron phosphate
OCVchr	open-circuit voltage before charge
OCVdch	open-circuit voltage before discharge
PSOC	partial state of charge
Rchr	charging resistance
Rdch	discharging resistance
SNL	Sandia National Laboratories
SOC	state of charge
Vmin	minimum operational voltage
Vmax	maximum operational voltage

ΔVchr   change in voltage on charge
ΔVdch   change in voltage on discharge
Wh   Watt-hour
W   Watt

1. INTRODUCTION

This work was supported by the U.S. Department of Energy (DOE) Office of Electricity Delivery & Energy Reliability. The DOE program goals are directed at supporting industry and utilities in the areas of:

- Development and evaluation of integrated electrical energy storage systems;
- Development of batteries, superconducting magnetic electrical energy storage (SMES), flywheels, super capacitors and other advanced energy storage devices;
- Improvement in multi-use power electronics, controls, and communications components;
- Analysis and comparison of technologies and applications; and
- Encouraging program participation by industry, academia, research organizations and regulatory agencies.

The work reported in this paper is part of our effort to characterize the performance parameters of advanced batteries. The iron phosphate Li-ion battery technology has recently entered the battery market with a number of manufacturers from the United States (*A123 and Valence*), Taiwan (*LiFeBatt*), and China (*AA Portable Power and K2 Energy*). This technology was originally developed and patented by Dr. John Goodenough and his team at the University of Texas in 1996, additional work by Université de Montreal (UDM) and Hydro-Quebec resulted in the addition of carbon to the Li-FePO4 in a process called carbon Nano-painting. This process improves energy density and power performance. In 2002, UDM and Hydro-Quebec, as owner or co-owner of many patents on LiFePO4 technology, established Phostech Lithium Inc. to commercialize the product. Phostech Lithium produces material to replace the LiCoO2 and other cathode materials typically used in Li-ion batteries. The LiFeBatt battery uses iron phosphate cathode material because it has a number of advantages over conventional Li-ion cathode materials. These advantages include lower cost, improved safety, and high power performance. The most significant limitation to the iron phosphate cathode material is the lower energy density at approximately 80 Wh/Kg vs. 165-180 Wh/Kg for $LiCoO_2$ and other oxide cathodes[1]. However, as a result of the safety and power performance improvements, the iron phosphate Li-ion battery technology is being evaluated for new market areas including hybrid electric vehicles, utility energy storage, and other large energy storage applications. In this paper the performance of the LiFeBatt cell was evaluated using high rate capacity tests, low temperature capacity, ohmic resistance, spectral impedance, partial state of charge (PSOC) pulse power cycling, pulse power performance, and an over-charge/voltage abuse test.

In Figure 1 is a photograph of one of the LiFeBatt 10 Ah cells. This cell has dimensions of 40 mm in dia by 139 mm long, weighs 360 g, and has a nominal rated capacity of 30 Wh. Table 1 summarizes the LiFeBatt cell specifications.

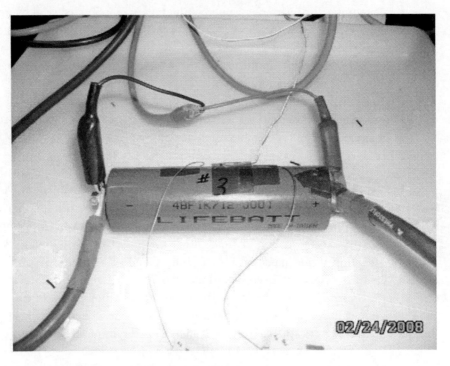

Figure 1. LiFeBatt Li-FePO4 Cell Under Test.

Table 1. LiFeBatt Cell Specifications

Model #PC40138F1W Cell	
Operating voltage window	2.10 – 3.65 V
Max voltage	3.65 V
Discharge end voltage (1C rate)	2.10 V
Charge Regulation Voltage	3.65 V
Full Charge Termination (V, I, Time)	3.65 V and 0.2 A or 60 min @ 3.65 V
Maximum Charge Current	30 A
Maximum Pulse Current, I	140 A
Maximum Constant Current, I	120 A
Internal Ohmic resistance, mOhm +25 °C	<6 mohm
Ah Capacity (0.25C and 1C rate)	10,000 mAh, 9,000 mAh
Energy stored in operating voltage window, Wh or kJ	80 Wh/kg 170 Wh/L
Overall dimensions, mm	171.1 x 40.6
Weight, kg	0.360
Operating temperature, °C	0 to 45 Chr -10 to 55 Dch
Storage temperature, °C	-10 to 45
Cycle life, cycles	1,500

2. TEST PROCEDURES

The cell test procedures used in this effort were initially developed as part of a plan to test batteries for partial state of charge (PSOC) pulsed cycling in utility applications, and the eight characterization tests used are itemized below.

1. **Capacity Test** – Establishes a capacity on each cell.
2. **DC Ohmic Resistance** – Establishes a resistance of the cell.
3. **AC Spectral Impedance** - Establishes the AC impedance of the cell.
4. **Cell Power Density and Specific Energy Density** – Measures the cell power and energy density.
5. **Cell Capacity and Recharge As A Function Of Temperature** – Capacity measurements at the 1C rate were conducted at 35°, 22°, 0°, -20°, -40 °C.

6. **Cell Utility PSOC Pulsed Cycle Test** – Measures the ability of the cell to PSOC cycle at high power for utility voltage support, frequency stabilization, and wind farm energy smoothing applications.
7. **Hybrid Pulse Power Test** – Measures the 10 second pulse power performance from 90% to 10% state of charge (SOC).
8. **Over Voltage/Charge Abuse Test** – Measures the effects of an uncontrolled continuous 10 A charge.

2.1. Capacity Test

The capacity test is used to determine the cell capacity and this test is done prior to testing to establish a baseline as well as being repeated at the end of PSOC cycle testing. This will help identify how the various tests in this plan affect the cell capacity.

The cell shall be tested for its capacity, as follows:

1. Each cell shall be charged at 0.5C (5.0 A) up to Charge Voltage (3.65V). On reaching the Charge Voltage, the current shall be allowed to taper while maintaining the Charge Voltage until the current tapers to 0.2 A. The ampere-hour input into the cell shall be measured.
2. The cell shall rest at open circuit for 30 minutes.
3. The cell shall be discharged at 0.5C (5.0 A) until the end voltage (2.1 V) is reached. The ampere-hour capacity of the cell shall be measured.
4. The cell shall rest at open circuit for 30 minutes.
5. Steps 1 to 5 shall be repeated 3 times. The third capacity measurement will be the recorded capacity.

2.2. DC Ohmic Resistance

The DC resistance shall be measured with a high current discharge pulse of at least $2C_1$ at the 100% SOC point. An oscilloscope will be used to measure the ohmic voltage drop and current ramp using the following steps:

1. Charge cell at 1C rate to the Charge Voltage (3.65V) and hold at voltage until the current tapers to 0.2 A.
2. Allow the cell to sit open-circuit for 30 min.

3. Measure the dynamic DC ohmic resistance of the cell using an oscilloscope by discharging at 2C1 for 2 seconds.

2.3. AC Spectral Impedance

AC spectral impedance measurements of the as-received and after PSOC pulsed cycling were also made. These measurements shall be made in the following manner:

1. Each cell shall be charged at 1C up to the Charge Voltage (3.65V) until the current tapers to 0.2A.
2. The cell shall rest at open circuit for 30 minutes.
3. The cell shall be discharged at the 1C rate to 50% SOC.
4. The cell shall rest at open circuit for at least one day prior to measurement.
5. Each terminal on the cell shall be fitted with two gold plated interconnects (to minimize contact impedance).
6. The cell shall be placed into an environmental chamber regulated at 25 °C for at least 12 h prior to initiating measurements.
7. All measurements shall be made with the cell in a controlled temperature maintained at 25 °C.
8. The cell shall be connected to the instrumentation in a four-wire configuration.
9. The peak-to-peak AC voltage shall be in a range to allow 1% accuracy of the impedance of the cell being measured, and in this case 3 mV was used.
10. The frequency range shall be large enough to encompass the anticipated network response, and in this case corresponds to a range of 100 kHz to 10^{-4} Hz. At least six different frequencies per decade shall be measured.
11. All measurements shall be made at 0 V vs. the open circuit voltage corresponding to 50% SOC.

2.4. Cell Power Density And Specific Energy

Measure cell capacity as close as possible to the following rates:

1	0.1C,	10 h
2	0.2C,	5 h
3	1C,	1 h
4	2C,	0.5 h
5	4C,	0.25 h
6	10C,	0.1 h

The cell capacity shall be measured as follows:

1. Each cell shall be charged at 1C up to Charge Voltage (3.65V). On reaching the Charge Voltage, the current shall be allowed to taper while maintaining the Charge Voltage until the current tapers to 0.2 A. The ampere-hour input into the cell shall be measured.
2. The cell shall rest at open circuit for 30 minutes.
3. The cell shall be discharged at specified rate until the end voltage (2.1 V) is reached. The ampere-hour and Watt-hour capacity of the cell shall be measured.
4. Steps 1 to 4 shall be repeated 3 times.

Using the data above, calculate the power and energy density with respect to volume and weight and display using the Ragone plot.

2.5. Cell Capacity and Recharge as a Function of Temperature

Cell capacity and charge and discharge characteristics at the 1C rate (10 A) were measured at 35, 22, 0, -20, and -40 °C. The capacity test procedure was the same as that described in the capacity measurement.

2.6. Utility PSOC Pulsed Cycle Test

The utility PSOC pulsed cycle test is designed to evaluate battery performance under short high power charge and discharge environments. In many utility applications the battery is required to both sink and source power for voltage support, frequency stabilization, and wind farm energy smoothing. In Figure 2 are actual utility data obtained from Charles Koontz of WPS Energy Services, Inc. showing the magnitude and duration of the power pulses required to support a utility application. In general, the pulse durations are minutes in length. The utility PSOC charge and discharge pulses chosen for this test are between 1.5 and 6 minutes in length at discharge rates between 1C₁ (10 A) and 4C₁ (40A). The goal of this testing is to evaluate PSOC pulsed cycling, cell stability, efficiency, power performance, thermal management, and charge management strategies.

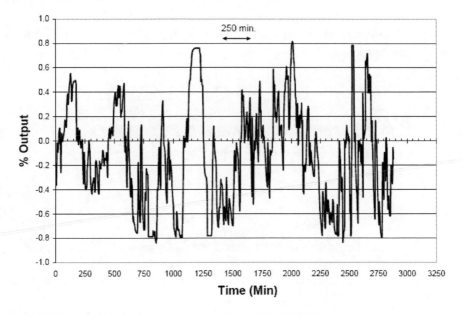

Figure 2. Typical Utility Energy Pulses (Charles Koontz, WPS).

The cycle profile in this test is illustrated in Figure 3 and consists of the following steps:

1. Charge cell at 1C rate until voltage reaches Charge voltage (3.65V).
2. Keep voltage at Charge voltage until current tapers to 0.2 A.
3. Rest for 30 min.
4. Discharge at 1C rate to end voltage (2.1 V).
5. Rest for 30 min.
6. Recharge cell as in step 2.
7. Discharge at 1C rate to 50% Ah capacity.
8. Rest for 5 min.
9. Discharge at 1C rate for 6 min. (10% DOD).
10. Rest for 5 min.

11. Charge at 1C rate for 6 min.
12. Rest for 5 min.
13. Repeat steps 9 through 12 for 100 cycles.
14. Measure available capacity as specified in steps 3 through 5.
15. Repeat 100 cycle profile five times per test sequence.
16. Evaluate battery performance and determine if higher power levels are possible and if the PSOC cycle interval can be extended to 500 or 1,000 cycles.
17. Additional testing may be conducted at the 2C, 3 min. or 4C, 1.5 min. rates and times (10% DOD).

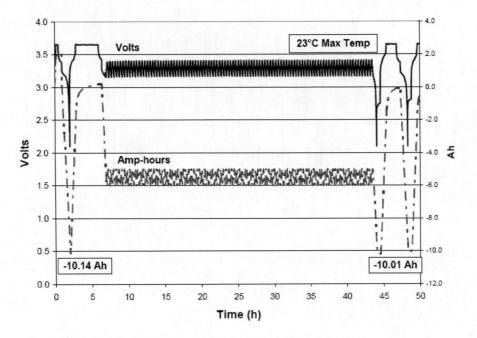

Figure 3. Utility PSOC Pulsed Cycle Test with 1C 6 min. (10% DOD) Charge/Discharge Pulses.

2.7. Hybrid Pulse Power Test

The Hybrid Pulse Power Test is extracted from the FreedomCAR Battery Test Manual For Power-Assist Hybrid Electric Vehicles. This test procedure uses a 10 second $5C_1$ discharge pulse and a $3.75C_1$ charge pulse 40 seconds apart (see Fig. 4). The test sequence is listed below:

1. Measure capacity at the 1C rate.
2. Fully recharge cell.
3. Allow cell to rest open-circuit for 1 h.
4. Discharge cell 10% at the 1C rate,
5. Allow the cell to rest for 1 h rest open-circuit (measure Voc).
6. Discharge cell at the 5C1 rate for 10 seconds (measure end of discharge V).
7. Allow the cell to rest open-circuit for 40 seconds (measure Voc).
8. Charge at the 3.75 C1 rate for 10 seconds (measure end of charge V).
9. Discharge at the 1C rate 10% of the cell capacity.
10. Repeat steps 4 through 8 until battery is at 10% SOC.
11. Record open-circuit voltage after the 1 h rest before the discharge pulse, record voltage at 10 second point in charge and discharge pulse and record open-circuit voltage at end of 40 second rest for each SOC.
12. Calculate discharge resistance using the 1 h open-circuit voltage and charge resistance using the 40 second open-circuit voltage for each SOC.

$$R_{Dch} = \frac{\Delta V_{Dch}}{\Delta I_{Dch}}$$

$$R_{Chr} = \frac{\Delta V_{Chr}}{\Delta I_{Chr}}$$

13. Calculate the Discharge Pulse Power Capability for each SOC using the minimum operational voltage.

$$Watts = V_{min} \bullet (OCV_{Dch} - V_{min}) \div R_{Dch}$$

14. Calculate the Charge Pulse Power Capability for each SOC using the maximum operational voltage.

$$Watts = V_{max} \bullet (V_{max} - OCV_{Chr}) \div R_{Chr}$$

15. Plot the discharge and charge power as a function of % SOC and discharged energy (Wh) at the 1 h rate.

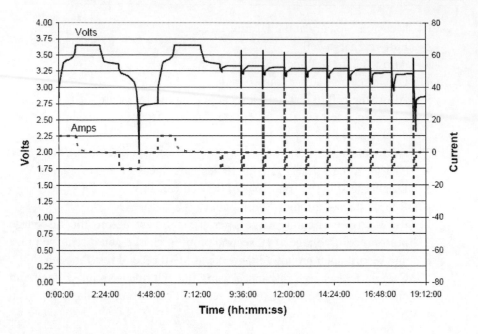

Figure 4. Hybrid Pulse Power Test.

Max power was calculated from the recorded data and using the equations shown in steps 13 and 14 above. Using ohms law, current is calculated by dividing the voltage by resistance. The voltage difference between the cell open-circuit and cell max and min voltages divided by resistance is assumed to be a measure of maximum current. In the second half of the two equations, the min and max voltages are multiplied by the calculated current to yield max 10 second power (Watts = V x I). Once the max charge and discharge power is calculated, then the power data can be plotted as a function of % SOC and battery energy level in Wh. The power vs. energy plot can then be used to scale an energy storage system for any power and energy requirement at specified pulse durations.

2.8. Over Voltage/Charge Abuse Test

The Over Voltage/Charge Abuse Test was duplicated from the LiFeBatt battery Product Specifications. In the Product Specifications the cell/battery is charged at the 1C rate (10 A) up to 12 V and 7 h. The test conducted for this report was modified slightly because of a number of unforeseen events. The charge and data acquisition was terminated when the cell under test lost all voltage and caused the data acquisition system and battery cycle test equipment to stop due to the loss in cell voltage.

3. TEST RESULTS

3.1. Initial and Final Capacity

Figure 5 shows the voltage and capacity (Ah) data for a discharge and charge on cell #1 as received and after 8,394 PSOC pulse cycles. At an initial capacity of 9.61 Ah this cell is within the capacity specifications of between 9 and 10 Ah. After 8,394 PSOC pulse cycles the capacity dropped to 8.91 Ah a loss of 7%. The final capacity test was able to recovered capacity from end of PSOC cycling measured at the 1C rate (8.61 Ah) to 8.91 Ah measured at the 0.5C1 rate. This recovery is attributed to the deep-cycle, lower rate, and full charges using an end of charge current of 0.2 A. In addition, the initial capacity measured at the 1C rate during the PSOC pulse cycling was 10.14 Ah. This increase from the initial value was attributed to an apparent "break-in" or formation period for the cell resulting from a number of deep-cycles before the PSOC pulse cycling.

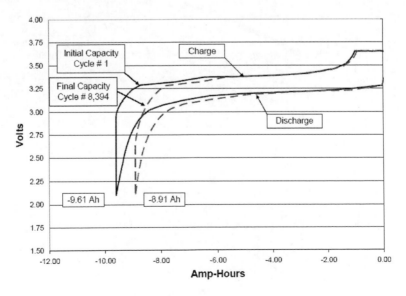

Figure 5. Initial and Final Capacity of Cell #1 (5A Chr/Dch).

3.2. DC Ohmic Resistance

Figure 6 shows the initial ohmic resistance measurement for cell #1 at 0.0036 ohms ($\Delta V/\Delta I$) and after 8,394 PSOC pulsed cycles at 0.0042 ohms. The measurement is made using a linear regression to obtain the average slope of the line. In this case the regression is a good fit to data. The slope of the current and voltage measurement is the average resistance value for a discharge between 0 and -20 A. This battery chemistry has a slope that is consistent throughout the discharge. In some battery technologies there can be a significant difference from beginning to end of the discharge pulse. The data in this measurement is significantly less than the manufacturer's specification of less than 0.006 ohms impedance. It is not clear how the manufacturer has measured the impedance. The final ohmic resistance measurement is also lower in voltage, but it is not clear why there was a drop in voltage at 100% SOC.

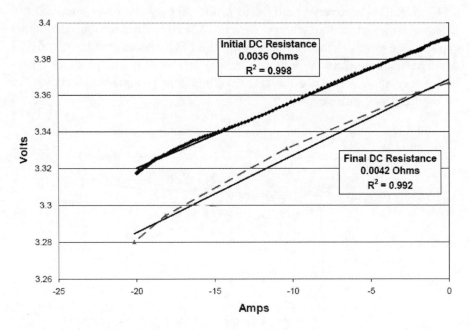

Figure 6. Ohmic Resistance Measurement On Cell #1 (100% SOC).

3.3. AC Spectral Impedance

A Nyquist plot of the impedance data before and after PSOC pulsed cycling is shown in Figure 7, and as seen there is very little difference in behavior. In fact, differences in the before and after PSOC pulsed cycling behavior cannot be observed when all of the data is displayed, and differences only become apparent after expanding the scale. The expanded scale high frequency response is also shown in Figure 7. The high frequency impedance of the cell is essentially the same before and after PSOC pulsed cycling. The small decrease in impedance from 4.01 m$_o$ to 3.74 m$_o$ after cycling may in fact be a reflection of the slight differences in SOC, as seen by the lower OCV of the cell at the time the measurements were made. Although we attempted to minimize this difference as described previously, the loss in cell capacity on cycling may result in different SOCs for the electrodes when the impedance measurements are made, and in the two electrode configuration we do not have a means for determining the relative contribution of each to the impedance measured. Also seen in the high frequency data is what is presumably a single passive layer having a total resistance of 1.52 m$_o$ before cycling. As in the case of the cell impedance, there is little or no change in the resistance of this film after cycling, and its resistance is estimated to be 1.6 m$_o$ after cycling. Again, because of the uncertainty in the measurements, this value is essentially unchanged.

These behaviors are significantly different from the more typical lithium-ion chemistries that operate at higher voltages, (eg LiCoO$_2$-based materials [2]). In these other cell chemistries we generally observe increases in all of the impedances measured, and in some cases the formation of another passive film. In the case of the LiFeBatt cell, we see essentially no change of impedance as a result of PSOC cycling. Based on this behavior, it would be very useful to complete an in depth study of the impedance behavior of these materials in order to identify the processes occurring and limiting conditions.

On the left is shown the full range of data collected, and on the right is an expanded scale version of the high frequency response near the intercept. The OCV of the cell at the time of the measurements was 3.3997 V and 3.295 V for the cell before and after PSOC cycling respectively.

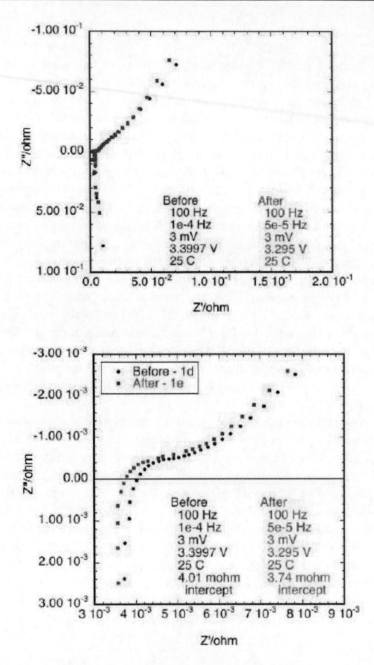

Figure 7. Impedance behavior of cell before and after PSOC cycling.

3.4. Cell Power and Specific Energy Density

In Figure 8 is a plot of the cell specific energy and power performance and in Figure 9 is the energy and power density using the Ragone plot. The results show a steep curve as the power level increased indicating minimal energy losses and a sharp rollover at high power as the cell discharge is terminated due to excessive temperatures over 60°C.

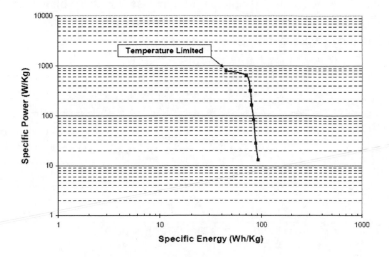

Figure 8. Specific Energy And Power Performance.

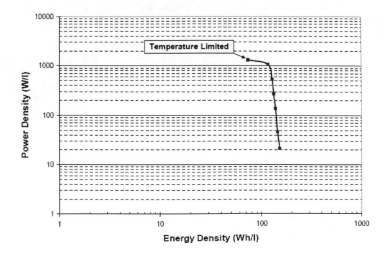

Figure 9. Energy Density And Power Density.

3.5. CELL CAPACITY AND RECHARGE
AS A FUNCTION OF TEMPERATURE

In Figure 10 are the 1C capacity measurements at 35, 25, 0, -20, -30, and -40 °C. The results show a rapid drop in capacity at -20 °C and below, and at -40 °C there is virtually no usable capacity. In Figure 11 are the recharge voltage profiles at the 1C rate. At -20 C and below, the recharge is dramatically slowed due to the rapid increase in cell voltage up to the regulation voltage. At these low temperatures the charge currents are very low requiring long recharge times.

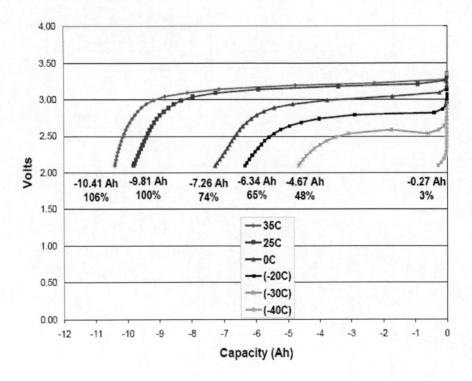

Figure 10. Capacity vs. Temperature at 10A.

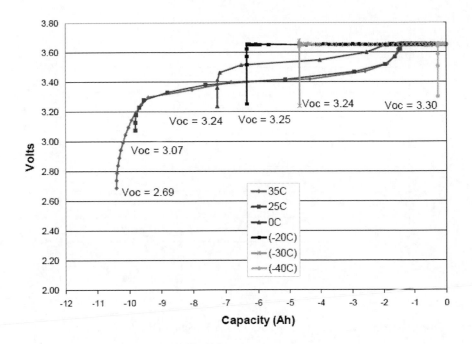

Figure 11. Recharge vs. Temperature at 10A.

3.6. Hybrid Pulse Power Test

The hybrid pulse power test results are shown in Figure 12 and 13. As expected, the maximum discharge pulse at a low state of charge is significantly reduced and as the SOC increases so does the pulse power, from 164 W (10% SOC) to 380 W (90% SOC). Usually the charge pulse has a similar decrease in power as the SOC increases, but in this case it remains fairly constant at about 200 W between 30 and 60% SOC. In addition, at 80 and 90% SOC the charge pulse is also constant at about 180 W. This behavior was replicated several times and might possibly reflect a functional dependence of cell impedance on SOC. In Figure 13 the available battery energy is plotted on the abscissa (X-axis) providing an operational range with available energy. This can be useful for scaling up to the necessary power and energy levels needed to meet the 10 second pulse power requirements of the desired system.

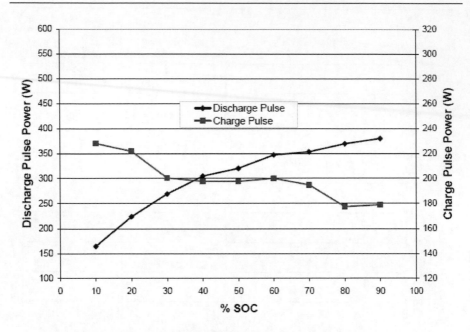

Figure 12. Hybrid Pulse Power Capability As A Function Of %SOC.

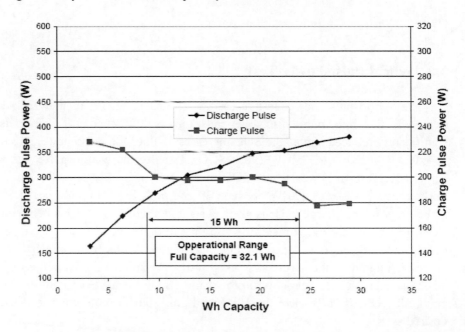

Figure 13. Hybrid Pulse Power Capability As A Function Of Capacity In Wh.

3.7. Utility PSOC Pulse Cycle-Life Test

In Figure 14 and 15 are the last Utility PSOC test results at 4C1 (±40 A) and the summary capacity plot of the 1C, 4C1, and 2C1 PSOC pulse cycling results.

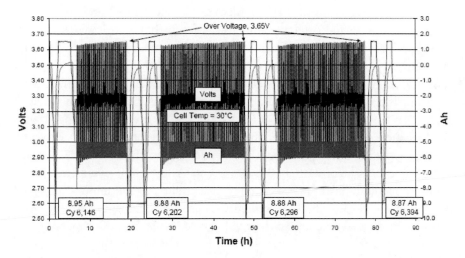

Figure 14. Utility PSOC Pulse Cycle-Life Test Results At ±40A for 1.5 min.

Figure 14 is a plot showing the steady increase in end of charge voltage at the 4C1 (±40 A) rate. In this plot, the cell voltage steps up to 3.65 V from 3.63 V in 56, 94, and 98 cycles, while at the same time the cell temperature increases from 22° to 30°C. Initially at cycle 500, the step up in voltage began at 3.58 V and required 917, 745, and 1,000 pulse cycles to reach the maximum end of charge voltage. As a result of the pulsed cycling, the ability of the cell to accept the 40 A pulse charge has degraded. Additional testing at 20 A starting at cycle 6,394 has also shown the same steady increase in end of charge voltage. Since the end of charge voltage at 20 A is well below the maximum limit, the cell could cycle well beyond 1,000 PSOC cycles without limiting current or implementing a recovery cycle to recover the voltage back to its initial value. In an actual application, this step up in voltage would limit the number and rate of the pulse charges available before a recovery cycle was implemented.

Figure 15 is made up of three different regions where the charge/discharge current and time was changed from 10 A for 6 min, to 40 A for 1.5 min, to 20 A for 3 min. These changes were the result of the initial characterization at 10

A, then high power testing at 40 A, and finally reduced power testing at 20 A. The plot shows that the capacity drops to about 88% of its initial value of 10.14 Ah after about 2,000 cycles. After 2,000 cycles, the capacity slowly fades down to 85% (8.61 Ah) of the initial value out past 8,394 cycles. At five points along the curve are 0.25 Ah increases in capacity at the same cycle number. This jump is the result of a second full charge after finishing a test sequence. The data shows that the recovery is short lived and the capacity quickly returns to the original trend line. It is difficult to know how many cycles are possible to the 80% capacity value, but based on a linear trend line, the cell may cycle for 18,000 cycles before reaching 80% of its initial capacity. Also, the spacing between capacity measurements at the high power 40 A rate is a function of how quickly the cell reaches the maximum voltage at 3.65 V. When the cell reaches 3.65 V, the PSOC cycling is stopped and a capacity measurement is initiated. In the 40 A test region, the curve shows that this is happening initially at about 1,000 cycle intervals but has decreased to near 100 cycles after the 6,000 cycle point. After 6,000 cycles, the cell is cycled at medium power (±20 A) and returns to the predetermined 1,000 cycle interval between capacity measurements. Thus, the cell is aging as seen by the accelerated increase in end of charge voltage and walk down in capacity.

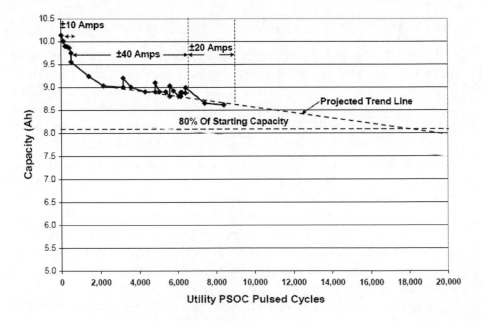

Figure 15. Utility PSOC Pulse Cycle Test Results.

3.8. Over Voltage/Charge Abuse Test

In Figure 16 the events in an over charge/voltage abuse test are documented. Initially, as expected, the cell voltage increases quickly while being charged at 10 A, but then slowly increases after 4.7 V. The cell voltage slowly increases for about 30 minutes while the cell temperature continues to slowly rise to about 100 °C at which time cell voltage spikes to the maximum value of 12 V. At about 110 °C the cell vents liquid electrolyte without any fire or sparks and then open-circuits at 116 °C. After open-circuiting and a loss of electrolyte, the cell looses all voltage at 120 °C. The data acquisition shuts down due to a no voltage condition, but temperature is manually monitored until the cell reaches its maximum value at 160 °C about 20 minutes after the cell open-circuited.

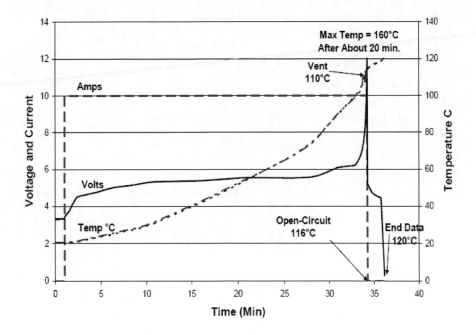

Figure 16. Over Voltage/Charge Abuse Test.

Table 2. LiFeBatt Summary Test Results

Test	Cell #1	Cell #2	Cell #3
Initial Capacity C/2 (Ah)	9.61	9.88	9.81
Initial DC Ohmic Resistance (Ohms)	0.0036	0.0037	0.0033
Initial AC Spectral Impedance ESR (Ohms)	0.0041	NA	NA
Cell Power and Energy Density at 10C1 Rate (W/l, Wh/l, W/Kg, Wh/Kg)	NA	NA	1,068, 116, 653, 71
Cell Capacity and Recharge As A Function Of Temperature (Ah @ 35, 25, 0, -20, -30, and -40C	NA	NA	10.41, 9.81, 7.26, 6.34, 4.67, 0.27
Cell Utility Cycle-Life PSOC Test (Cycles, % Capacity Loss)	8,394 / ~15%	NA	NA
Final DC Ohmic Resistance and Spectral Impedance (Ohms, ESR)	0.0042	NA	NA
Final AC Spectral Impedance ESR (Ohms)	0.0037		
Final Capacity Test C/2 (Ah)	8.91	NA	NA
Hybrid Pulse Power Test (Charge/Discharge Power (W) at 50% SOC)	NA	NA	198, 331
Over Voltage/Charge Abuse Test (Vent Temp, Max Temp, Fire Y/N)	NA	110°C, 160°C, N	NA

4. SUMMARY

Table 2 provides a brief summary of the test results on three cells. After capacity and impedance measurements cell #1 was PSOC pulsed cycled for 8,394 cycles (see Table 2). The other two cells were characterized for over voltage/charge, cell power, and capacity vs. temperature, as indicated in Table 2. These results indicate that the LiFeBatt cells meet the manufacturers specifications in capacity, internal ohmic resistance, max power, and specific energy. The temperature performance at 35, 25, 0, -20, -30, -40 °C show that both discharge capacity and recharge voltage are significantly affected by low temperature especially at temperatures below -20 °C. From -20 °C to 35 °C the capacity will increase by about 0.8% per °C. At temperatures below -20 °C the capacity drops dramatically to just 3% at -40 °C. The 10 second pulse power capability values can be used to scale up for larger pulse power applications and show power levels of 325 W/cell on discharge and 300 W/cell on charge at 50% SOC. The utility PSOC pulse cycle test results after 8,394 cycles suggests that the LiFeBatt cells will PSOC pulse cycle up to the 4C1 rate for utility applications. Performance degradation was identified by capacity fade (10 to 15% loss - based on 9.61 or 10.14 Ah initial capacity) and by the increase in end of charge voltage during high power and medium power pulses. Based on the slow trend of capacity fade, the test cell should PSOC pulse cycle well beyond the 8,394 cycles tested before reaching 80% of initial capacity. This assumes that there will be no premature failure mechanisms that terminate life early. The ohmic resistance measurements and spectral impedance measurements before and after the PSOC pulse cycling have indicated only a slight increase in ohmic resistance and a very slight drop in ESR for spectral impedance. The ohmic value increased from 3.6 to 4.2 mohms, while the AC spectral impedance ESR value decreased from 4.01 to 3.74 mohms. This is a very minimal change, if any, and indicates no similar degradation as seen in the $LiCoO_2$ materials where increases in all of the impedances measured occur, and in some cases another passive film can form. Finally, the over charge/voltage abuse test indicated that the LiFeBatt cell can fail without fire, or damage to other external systems if those systems can handle the 160 °C max temperature and electrolyte venting.

5. REFERENCES

[1]   Handbook Of Batteries, 3th edition, G. W. Linden, McGraw-Hill Handbooks 1995, ISBN 0-07-135978-8,

[2]   D. Zhang, et al, "Studies on capacity fade of lithium-ion batteries," Journal of Power Sources, 91, (2000), pp.122-129.

INDEX

T

Taiwan, vii, 3, 32, 33
technologies, 3, 14, 33, 44
technology, vii, viii, 1, 3, 31, 33
temperature, vii, viii, 1, 3, 4, 6, 7, 24, 27, 30, 31, 34, 35, 37, 51, 53, 54
test procedure, 5, 8, 10, 35, 38, 40
testing, 2, 5, 6, 8, 16, 17, 32, 36, 39, 40, 51, 52

U

United States, vii, 3, 33
USA, 32
utility energy storage, vii, 3, 34
utility PSOC cycling, 30

V

vehicles, 34

W

wind farm, vii, viii, 1, 5, 8, 31, 36, 38

X

X-axis, 22, 49

Y

yield, 11, 42